I0004582

Practical Guide to
Project Scope Management

SHYAMKUMAR NARAYANA

Copyright © 2018 Shyamkumar Narayana
All rights reserved.

ISBN: 1-4392-1798-X
ISBN-13: 9781439217986

Visit www.booksurge.com to order additional copies.

To My Father and Mother for their unconditional love and support and to my family.

Acknowledgments

I want to thank God for giving me the practical experience and wisdom which enabled me to come up with this book. I also specially thank the Project Management Institute for allowing me to use a definition from Project Management Body of Knowledge called PMBOK in my book. Special thanks to Booksurge for providing me the opportunity to publish this book and also for the constructive and encouraging feedback I received from the editor of Booksurge.

Preface

··

    I wanted to share my experiences on how to manage Project scope through my book "Practical Guide to Project Scope Management". The readers of my book should find valuable and practical insights to Project Scope Management.

Chapter 1
Introduction

What is project scope management?

According to PMBOK (Project Management Body of Knowledge), "Project Scope Management includes the processes required to ensure that the project includes all the work required, and only the work required, to complete the project successfully. It is primarily concerned with defining and controlling what is or is not included in the project."[1]

Project scope creep is a critical issue in scope management. On one hand there are organizations in which projects are managed without a formal methodology. There is no rigor of managing changes in scope. Any changes in requirements communicated by the business are directly incorporated into code without evaluating the effect on the original timeline and cost. This results in schedule changes and cost overruns.

For example, imagine you have a twelve-month project worth $1 million. If your organization is not strict in managing scope creep then eventually you will be managing a project that will cost more than $1 million and will stretch beyond twelve months. If you have a commitment with a customer for a project delivery date, you will end up having a dispute with the customer who will refer to the original contract regarding the project delivery date. The cost of the work that extends beyond twelve months will not be borne by your customer, which adds to your cost. In the end you've lost money and your customer is unhappy.

On the other hand, take, for example, the same million dollars and a yearlong project managed by an organization with proper scope management practices in place. This organization will receive in writing any changes mandated by its customers and will pass the request for change with an impact analysis to the change control board and inform the customer what it will cost in terms of time and money to incorporate the change. Only with the customer's agreement to buy into the extra cost and extension of the project end date, if required, will the changes be approved for incorporation. If the customer comes back tomorrow and complains about the extra cost incurred and the extra time taken, then the manager of the project can show the valid documentary evidence of the change approval.

[1]Source: Project Management Institute, A Guide to the Project Management Body of Knowledge (PMBOK® Guide) – 1996 Edition, Project Management Institute, Inc., (1996). Copyright and all rights reserved. Material from this publication has been reproduced with the permission of PMI.

Hypothetical Example:

Type of Scope Management	Original Schedule	Original Cost	Revised Schedule
Weak Scope Management	12 Months	$1 Million	13 Months
Strong Scope Management	12 Months	$1 Million	13 Months

Hypothetical Example (continued)

Type of Scope Management	Revised Cost	Who Bears the Cost
Weak Scope Management	$1.5 Million	Performing organization bears five hundred thousand dollars in extra cost.
Strong Scope Management	$1.5 Million	Customer bears five hundred thousand dollars in extra cost.

Schedule and Cost Variance:

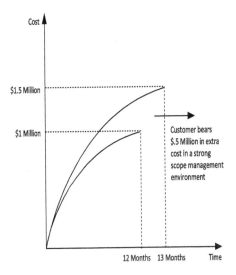

Chapter 2
Factors required for proper scope management

What are the factors required for a proper scope management process to be in place in an organization?

1. A scope management plan that will define how scope will be managed in the project. This will link to the overall scope management policy of the performing or client organization.

2. A change control board (CCB) consisting of members who will emphasizethechangecontrolprocessorpolicethechangecontrol in the organization. The CCB will be the approver of changes to projects. This board will scrutinize the changes via the impact analysis on the original timeline and cost. They will prioritize the change requests coming into the organization based on:

I. Nice to have

For example the user can have a color background in an order entry screen. The point is that the user can proceed with his or her work with the existing black and white background order entry screen.

II. Can live without it

The user can have a spell-check option in the memo text field in a data entry screen. The user can live without it, as the spell-check can be done using commercially available spell-check software.

III. The system will break without it

For example, an account number entered on a data entry screen needs a validation check to determine if the account is an existing account. If this validation is not there the system will function erroneously, i.e., it will give unpredictable results.

IV. Government or regulatory requirement

A bank has to create reports for financial analysis and report quarterly and yearly financial performance to the government for appropriate audit on the operations of the business.

The CCB will approve the change request based on the set priority and the impact to the project, the organization, and the customer. They will also obtain the necessary funding to accommodate the change after discussing it with the senior executives of the performing organization and the customer's executive representatives.

3. Discussion with project team, business, and systems personnel includes notations in requirements documents. Appropriate sign-off from key stakeholders is obtained and captured in the document. Once sign-off is obtained, a baseline is established. Any changes to the requirements documents will be carried out after obtaining approval from the CCB.

4. The organization should have a document management system that emphasizes version control. During audit of the project, the auditor can verify whether key or baselined documents were altered without going through the change control process. This will be a red flag to alert higher authorities that there was a slip in following scope management process. Corrective action should be taken immediately to bring the project on track. This is true to the saying: "A stitch in time saves nine."

5. have a list of key stakeholders for your project. The stakeholders will come from your customer organization, the performing organization, and your business organization. Stakeholders can also include your vendors. It is important to

have a good rapport with your stakeholders in order to get them buying into the progress of your project. A Typical list of stakeholders would be:

A. Project Manager

B. Program Manager

C. Director of IT or head of the IT Department

D. Business Manager or Manager of Business Group

E. Client Business representative

F. Change Control Board

The stakeholders have to be kept abreast on the progress of the project. To facilitate this communication, various metrics can be generated for the project and reported to key stakeholders. One example would be to create the project performance report about schedule and cost variance, i.e., the earned value analysis report. Another example would be to report on the number of defects discovered during uAT (user Acceptance Test) and how many of those defects were resolved in a timely manner, etc. The stakeholders also have to be in the loop on any project or project team issues or concerns in order to resolve them in a timely manner.

6. have a list of key deliverables required for a project and constantly monitor whether the key deliverables are being efficiently prepared and baselined to avoid last-minute surprises. Verify whether the code that is developed matches the requirements documents. Oftentimes you can see developers in discussion with business subject matter experts changing code on the fly. The requirements analyst and the project

manager (PM) are not in the loop during the discussion. This can be a disaster for the project. The feeling is like an ant slowly eating your candy without your getting a chance to taste the candy. Silently your project budget, in terms of time and cost, are eaten up and you as a project manager are not aware of it. Put all controls in place to eliminate this situation. An effective solution would be to have the concept of phase gate or phase exit applied in your project, i.e., at each phase exit point compare what has been delivered with what is required in the phase exit checklist and report variations for timely action by the project team members and the business if necessary.

"A Stitch in Time Saves Nine"

"The feeling is like an ant slowly eating your candy without your getting a chance to taste the candy."

Chapter 3
Situations where a scope creep can happen

Although a few were listed earlier, some key situations are highlighted below:

1. A developer has written code and has a meeting with the business SME (Subject Matter Expert) on a screen design. The SME suggests a few changes that were not in the original design. The developer, having good rapport with the business SME, modifies the code without informing the project manager or the requirements analyst and consumes a day more to complete the coding activity. This eats up the project budget.

The project manager should alert the project team of such a situation and encourage the project team members to keep the project manager in the loop in such situations. The project manager should educate the business SME or user to follow the change control process, i.e., to submit a change request through the proper channel for approval.

2. you as a project manager are in a strong matrix environment. your boss and the client representative have a good relationship. They discuss scope increase or changes without your presence in the discussions. Developers make changes to code without your knowledge and the knowledge of the requirements analyst based on your boss's directives. This creates schedule and cost overruns. As a PM without authority you are however responsible for bringing order to the situation. you will have to inform your boss about the importance of the change control process so that scope creep can be managed effectively. Get your boss and client representative in a room and come to an agreement that all scope changes will be communicated to the PM and the CCB, i.e., the changes will be routed through the proper communication channel.

3. The requirements analyst discusses requirements with the business user and comes to an agreement on the requirements. The developer starts to code based on the design in the requirements documents. however in a subsequent meeting with the user, the user realizes that more features need to be added and the requirements document is modified without consulting the PM. This translates to more coding time and it affects the project budget.

The project manager should emphasize to the business user and the requirements analyst that changes to requirements should be routed via the change control board, i.e., the change control process should be followed. Emphasis should be focused on baselining project deliverables in order to manage scope changes.

4. During system test, the tester finds navigating through a newly developed screen difficult. The same is discussed with the business user who suggests a few changes. The tester conveys these changes to the developer and the code is modified accordingly. The project manager is not in the loop here. This situation affects project timeline and cost, which the project manager is accountable for.

The project manager should highlight the importance of the change control process to the tester, developer, and again the business user. This will keep the project changes in check.

5. During the user acceptance test, the business user finds certain options missing in a newly built screen. The user suggests that the requirements analyst revise the design document. The project manager is not in the loop here. The design documents, the code, and test plans are modified to suit the business user's

needs. This becomes very costly for the project. The cost of the project increases and also the project end date is stretched now.

The project manager in this situation should again emphasize the importance of the change control process to key stakeholders in order to appropriately manage scope changes.

Chapter 4
Examples

Example 1 of an Increase in Project Scope

Say for example you are managing a manufacturing project wherein the outcome of the project is to print labels for Toys. Over the course of the project, your stakeholders discover that the labels have to be printed on a 3D printer vs. a Laser printer. At this discovery, you must change the scope of the project to accommodate the printing of labels on a 3D printer. However you have to go through a change control process for getting the buying from your project sponsor and the key stakeholders on the new scope along with the new timeline and the updated project cost. This step is very critical for the project in order to avoid any surprises later on in the project.

Example 2 of an Decrease in Project Scope

Another example is decrease in scope. You are managing a finance project where in the project is to write a software for a new government regulation around interest rates which has 10 business rules. Later on in the project, your stakeholders realize that not all 10 business rules are required for go live and they arrive at this decision due to the organizational budget limitations. This results in the reduction in project scope. You have to go through the change control process here also to get your project sponsor and the key stakeholders approve the reduction in scope along with the possible reduction in project cost and a shorter project timeline. Also your project resources could be released from the project sooner than planned.

Example 3 of an Elimination of Project Scope

The third example is the elimination of the project due to a newer technology making the current project obsolete as the organization cannot compete in the market with the older technology. This discovery happens over the course of the project. This can also lead to a totally new project scope using the newer technology and as warranted by the organization leaders. In either of the scenarios, the project manager should get the appropriate sign off from the project sponsor and the key stakeholders on the future direction i.e. it could be to either officially close the existing project and release the project resources or it could be to create a new scope outlining the new timeline and cost. In either case, you must get the buying from the project sponsor and the key stakeholders.

Chapter 5
Exercises

Exercise 1

Say for example you are going to manage a transportation system implementation project. You are new to the organization and team. You observe that no meeting minutes are documented after a meeting, there is no formal project plan in existence and a roadmap to complete the project also doesn't exist. What do you do then to fix the problem?

Exercise 2

You oversee a project that is executed using an agile methodology. You notice that the senior level executive within your organization has good relationship with the client executive and accommodates changes to the project scope midstream into the project. You are not aware of it or are not briefed about it and you see team members are demoralized and the project schedule is slipping. How do you approach the situation and fix the issue and bring the project on track in terms of scope, budget and schedule?

Exercise 3

You are tasked to manage an insurance project. You are dealing with SME's who have been with the company for a long time and are highly political. The SME's show a lot of resistance to formal project management methodology and processes. How do you bring things to order winning over the SME's in following a formal project management process for the ultimate success of the project and to the benefit of the client organization?

Chapter 6
Conclusion

As explained earlier, project scope creep is a critical issue in scope management. As a project manager, you can manage scope creep well by implementing a well-thought-out change control process in your organization. you can educate your project team and your business community on sound scope management practices in order to eliminate scope creep and keep your project budget under control. This will ensure peace of mind for the PM and the stakeholders.

www.ingramcontent.com/pod-product-compliance
Lightning Source LLC
Chambersburg PA
CBHW060937050326
40689CB00013B/3123